Help yourself - can you use the science method big bang to change your life?

By Alan Kennedy

To my wife Hannah for her love and support

Table of Contents

Chapter 1 Research Proposal - Your Big Bang Plan to succeed, can you go from nothing to creating Your universe?

Future scientist, your welcome to the world of success. Are you ready for your big bang? If yes get yourself ready to go from nothing to everything and enjoy the experiment of a lifetime. Here you learn scientific steps to transform your life, future and everything around you.

Before going further, answer one question - Why are you here? Answer this right, and then you are well on your way to success. Now below and for the rest of this book I will provide you with how but only you can and must know the why.

This book will be your lab book for the rest of your life. You must be ready to take in the key concepts and use your senses to observe and record your findings. This will be the key to your lifelong success

If you made it this far, congratulations you have taken the first and most important step starting, but like all good scientists you have to first complete the training.

So what is this training? It is not the physical training, unlike that in a physical lab but here is mind training. Here you will go from inexperienced self into the fantastic scientist of the mind that is lurking just below the surface.

Scientific Paper

First this is where you discover the main quality that all good scientific papers require before they even get started.

A Great Idea, Ability? Or perhaps a fantastic new insight into something never thought of before?

All very good ideas but it is none of the above. In fact the first thing that a scientist requires is...........A RESEARCH proposal plan.

Now this requires a well-structured amazing plan to achieve your idea. So what you must do first is create a plan to succeed.

Insert Image here of plan

Now I hear you say, that's all very well and good but how do you do that. After all planning stuff, it's so tedious and all. But this is in fact the most important, as well as the toughest part. Without a good plan or research proposal, you will not get very far. After all how do you get anywhere if you don't know what you're doing, where you're going, what successful experiment will look like, how to conduct it or what the results will be, never mind even getting anywhere near succeeding?

Plan

So this is the first part - to create your Research proposal. This can be whatever you wish but I will use the piano example below as an example. However you can insert whatever research proposal or goal you want, just make sure to use the proposal below.

A good proposal or plan will contain the following elements - Take Note of below as this is the first step to becoming the great scientist you know you can be. Below is the example of someone that wants to learn piano and how they will do it.

1. *Have a straightforward, descriptive, and informative title*

Eg I want to learn piano. I will do this by taking piano lessons and this will show I can achieve the ability to play music I enjoy on it proving I have learnt the piano and can play it.

So you must know exactly what you want, how to do it and a clear account of exactly what the question is that your research will address

2. An account of why this question is important and worth investigating

Eg I want to play piano to impress friends. This is a good reason, but better it is better if the reason can be intrinsic eg I want to play piano to improve my skills so I can pass on these skills to my children when older. The reason is that you will be more motivated to do it, if doing it for yourself.

3. An assessment of how your own research will engage with recent study in the subject

Eg I will research people in the area that can tutor me; I will research online about piano music I can get and also techniques that others have used previously to learn this skill.

4. A brief account of the methodology and approach you will take

Eg I will call up someone who is willing and able to tutor me at the times I can do. I will practice in between to improve upon areas that I need to achieve this goal.

5. A discussion of the primary sources that your research will draw upon an indicative bibliography of secondary sources that you have already consulted and/or are planning to consult

Eg I will record what I did to succeed here, the steps I took and sources I used as this will make achieving other goals far easier in the future

Well done for completing all of the above, whatever you goal is or was you are far more likely now to achieve it as you have overcome the most difficult step - Beginning.

Chapter 2 Abstract - Plan your big bang strategy?

Well congratulations, you now have a research proposal plan to support a big bang strategy.

Well now you have a plan. The next step that all good scientific papers require is an abstract or overview of what is happening. This is important as this will provide insight into what is going on and your strategy and key words or important parts here. Let's look this over below

Here the successful scientist must draw out a strategy of how to achieve the goals set out beforehand. So for example with your paper goal now been accepted you can now analyse what strategy to use here and the key ideas that will help you get there

Here is an abstract example from the previous mentioned task of learning the piano

"The goal of this proposal was to learn how to play the piano. Here I now undertook the proposals in my plan. I contacted a local music tutor who was able to tutor me once a week for 1 hour. I now acquired a piano and got some music books which I now practiced every day for at least one hour. The result was that I was able to make major improvements in my piano learning, as noted by my tutor. I was able to complete several songs and improved markedly over time. The results show that making an effective plan, carrying it out and implementing it as well as revising work I was struggling in meant I was able to achieve the goal of learning the piano. This method can be used in future to achieve other goals, perhaps learning a new language or travelling to a new country."

What you can see above is that the proposal is written talking about what going to do, the results of what you did and finally what they mean.

You must also envision yourself as carrying out all of the stages as an effective scientist and to be able to have the right equipment and ideas, to get the results and most importantly analyse them and see what else they can be used for or to fix problems that arose before.

Well now you can create an effective abstract you are now ready to begin.

Chapter 3 Scientific Introduction and Aim (Beginning and goal set) - Get Ready, the big bang begins?

Well now young scientist is where you will get ready to start.

So an Introduction and aim - This is where you decide what you are going to do and create an aim.

Introduction

Well this step is to cover some of the background of the topic. So continuing in the piano metaphor, you need to do some scientific research.

"What was I studying? Why was it an important question? What did we know about it before I did this study? How will this study advance our knowledge?. The answers to these questions will help introduce your topic and provide a background that you can use.

Also in a scientific paper you would use references eg primary scientific papers, which it is very important to reference correctly. So now it's time for you to go onto the next section. So what are you aiming for - only you can know this but an example of the piano aim is below.

Aim

The null hypothesis is that using these techniques will have no effect on my learning the piano.

What you say???.

Well thought this would get your attention here in case you were sleeping. Now you might be wondering, what on earth is going on? Surely the whole point of this is to succeed. Well young scientist, you need to pay more attention here.

The aim is not to show something happen. It
is to disprove the current theory that you're
doing stuff will have no effect by showing
that change happens when you change things.

The reason for this is that in science, it is
very difficult to prove something did happen
as it could be a million different reasons.
Eg maybe your hard work did help but how do
you know it was not just your ability or
luck. It is very important to show that it is
in fact your actions that lead to the change.
As without this, you could grow lazy and
believe that you inherently can just get
things and not that you have to work hard for
them.

Introduction and aim
Now you have and introduction and an aim, you
can put them together and see how they look
below

Learning the piano is a tough subject as shown from those that learnt it before. For example it is said that can take hundreds of hours of practice to become an expert. However the benefits of this are well known, with physical changes seen in the brain of those that play the piano. Also this talent can spark great things in those that learn it, especially when young still".

Here my aim is to disprove the null hypothesis that using these techniques will have no effect on my learning the piano.

As you can see above, this combines the introduction talking about background, history and reasons to learn as well as benefits of the work being done with the null hypothesis or aim. This is similar to that of a good scientific paper and is one good way to show that you can achieve your goal.

So whatever you aim is, create a null hypothesis that the actions will not affect it and then you can move onto the next stage.

Chapter 4 Methods and materials (What you do and what you need to do it) - The Big bang gets going?

Well congratulations young scientist you now have a Research proposal created, an abstract of the work and now have an introduction and aim done.

Then next step is to find, what we need to do what we are going to do.

Now you're going on back to look at the previous example of the piano.
We have created the introduction of the topic and null hypothesis that what we are going to do will have no effect on piano learning.

So this is where you gather everything from the previous plan together.   So you begin work to contact piano tutors and arrange the time for them to come out. You also begin working on the practice based on the schedule

you planned previously and locate the music
books as shown below.

Method and Materials

Here is where you begin looking at what you do
and what you need to get it done. It is
important to record this area as accurately as
possible so that in the future you can review
it. This allows you to make changes if you
want to use the same technique for other
goals.

<u>Example</u>

I will use the piano example here to show what
needs to be recorded

1.   Contact and arrange tutor
2.   Order or buy in all the materials needed
eg piano books, piano, learning piano apps
etc.
3.   Practice at the times set in the plan -
how long, what times

Here you record as much detail as possible.
This makes it far more likely you will be
successful as sometimes the slightest detail
is crucial in succeeding. Here you will
record who you contacted and how. Then
organise the time for them to come on out.

This is important, especially if something
goes wrong, for example if poor tutoring,
perhaps it was the location you found them
that was the problem. If that was the case
you could go back and redo the experiment and
succeed next time. Also if the tutor is fine,
perhaps it was the times so you can change
that here as well.

This is a vital stage as you can see, without
implementing the plans made previously,
nothing will happen here at all. This is also
useful in case you want to replicate your work
elsewhere eg If you wanted to go to the gym,
fix your finances etc. By recording this it

allows you to go back and improve or use it to improve yourself elsewhere. Here only record information that is relevant eg do not mention about unimportant details, but do cover potential problems etc.

Method and materials example

1. Contact and arrange tutor.

I contacted and got tutor to come out. They agreed to do 1 hour time slots every Thursday. One week they were on holiday so not able to make it but otherwise the experiment remained valid.

2. Materials required.

I purchased the following materials - 2 piano books, piano from gumtree and downloaded 2 piano apps as well. I also did research on the internet to find out more about pianos and how to prepare myself.

3. Practice

As well as the 1 hour time slot daily, I also arranged to practice for 1 hour every day as well, I used afternoons on weekends to fit in with school/work as well as evening during the week.

Chapter 5 Results (Self-explanatory here really) - The Big bang Result?

Results

Well this is quite obvious; this is the reason why you did all the work in the first place. Without any results you would have nothing to work from here, would you?

This requires you check what you do over time and factor in what you did

Eg Create a way to record what you're doing, otherwise how do you know if your succeeding.

Again it is important to record your results, so at the very least you know how you have done even if unsuccessful. In fact sometimes the best results are those that appear "Unsuccessful", when really they are just unexpected. There is a big difference between the two, in fact is dubious that any result is unsuccessful and I would argue that the only result you really get is called feedback.

Example

Result

Below are the results for one month

Day	Tutor	Book 1	Book 2	Piano	Practice	Improved
1	Yes	Yes	No	Yes	Yes	No
2	Yes	Yes	No	Yes	Yes	No
3	Yes	Yes	No	Yes	Yes	No
4	Yes	Yes	No	Yes	Yes	No
5	Yes	Yes	No	Yes	Yes	No
6	Yes	Yes	No	Yes	Yes	No
7	Yes	No	Yes	Yes	Yes	Yes
8	No	No	Yes	Yes	Yes	Yes
9	No	No	Yes	Yes	Yes	Yes
10	No	No	Yes	Yes	Yes	Yes
11	No	No	Yes	Yes	Yes	Yes
12	No	No	Yes	Yes	Yes	Yes

13	No	No	Yes	Yes	Yes	Yes
14	No	No	Yes	Yes	Yes	Yes
15	Yes	No	Yes	Yes	Yes	Yes
16	Yes	No	Yes	Yes	Yes	Yes
17	Yes	No	Yes	Yes	Yes	Yes
18	Yes	No	Yes	Yes	Yes	Yes
19	Yes	No	Yes	Yes	Yes	Yes
20	Yes	No	Yes	Yes	Yes	Yes
21	Yes	No	Yes	No	Yes	Yes
22	Yes	No	Yes	No	Yes	Yes
23	Yes	No	Yes	No	Yes	Yes
24	Yes	No	Yes	Yes	No	Yes
25	Yes	No	Yes	Yes	No	Yes
26	Yes	No	Yes	Yes	No	Yes
27	Yes	No	Yes	Yes	Yes	Yes
28	Yes	No	Yes	Yes	Yes	Yes
29	Yes	No	Yes	Yes	Yes	Yes
30	Yes	No	Yes	Yes	Yes	Yes

1. Tutor

The work with tutor, as shown in the table showed that one completed 1 month successfully. The only one not successful was when the tutor was on holiday. The result showed that the work was completed and had a strong correlation with improvement.

2. Materials required.

The work for the 1 week showed that the 1st book was inadequate. The tutor recommended that I use the 2nd book instead and as can be seen the learning of songs was far more successful when the 2nd book was used. The piano remained the same each week; it required tuning after 3 weeks for a few days so out of use but after that it was excellent

3. Practice

As can be seen in the graph above the practice was very successfully carried out, except for a blip in the 3nd week when I developed a

cold while the piano was out of use. The
rest of the time was fine and an extra
practice was carried out when the tutor was
away.

The above shows the importance of recording
the data from what you do and also making sure
that you actually do what you say you will.
In scientific papers a lot of data analysis
will go into very complicated experiments but
it all is dependent on the right data being
recorded in the first place.

Chapter 6 Discussion (What happened and why, how to change or insights for future) - Big Bang what's next?

So what happened, why and how you can change or what you can do for the future?

This is the important part, where you discuss what happened, after all if you do not know what happened as the old saying goes you are doomed to repeat it.

So how do you avoid this problem in science? Here we look at the results and try to cover as much as possible. In fact this is one of the reasons for having a null hypothesis as trying to prove something has actively changed is far more difficult than proving that it did not happen.

So what you can do here is assess whether you were successful or not, what results you got and any ideas or improvements you can do in

the future or what you can apply this too in
the future. Below we will use our piano
learning one as an example below.

Discussion

In the results you can see that the learning
was significantly improved by using a tutor,
as the one week the tutor was away the
learning did dip slightly. It is likely that
it would have fallen further if more practice
had not been done on the same week. Also
issues with the piano did seem to affect
learning as before tuning it did seem to have
issues but afterwards it was excellent for
learning. Also it can be seen that poor
quality learning materials had an effect as
better learning did take place when using
different piano learning book. As you can see
also that practice makes perfect does seem to
be true, with poor results when no practice
was done compared to that when practice was
being done regularly. This seems to show

that for other areas learning is improved when you have an expert tutor or someone that has done it before to help, good quality equipment to assist you and also practice, practice and practice.

Overall the null hypothesis - null hypothesis that using these techniques will have no effect on my learning the piano has been disproven.

Chapter 7 References (Your helpers, inspiration) - Thank you all and good night - Till the next big bang?

Finally you discussed the results you got and finished.

Well not quite, one final step after all your fantastic success is. The next task is to thank people for their help by creating a bibliography or reference list.

Why do this. This step is just a way of acknowledging help that you have had. However it is more than that, it means in the future if you wanted to find help you know where to go and how you got here.

In the end the only way to know where you are going is to see where you have been.

For example in the previous part we covered it would be the tutor, books you found online etc.

References example

Piano from gumtree

Tutor

Piano books

Piano apps

Well you are now ready, the scientific method is here for you. Go forth and bring your own big bang to the world.

www.ingramcontent.com/pod-product-compliance
Lightning Source LLC
Chambersburg PA
CBHW020714180526
45163CB00008B/3077